THE CORRECT MILITARY STRATEGY FOR THE VIETNAM WAR

(And Many of your other Military Historical Questions Answered)

By

Merrell D. Wils..

Printed in the United States of America

First Edition

ISBN 978-1-7325851-3-3

Dedication Page

To Everyone in my Extended Family Past, Present, and Future.

PREFACE

Why This Book?

Everything should have a reason. This book is written first to show what the correct military strategy of the Vietnam War should have been. Secondly; it was written to counteract the defeatist attitude of those against the war to preserve a Democratic Republic of South Vietnam and the anti-war demonstrators of the period; who claimed that there was no way to win the war. Thirdly; it was written to show that at least one person....myself; knew what the correct strategy to win the war should have been at the time....as I knew this in 1967.

The successful prosecution of the war would have meant the establishment of a Democratic Republic of South Vietnam; much as the Korean War led to the establishment of a Democratic Republic in South Korea. This is not to say there will never be a reunification of the Koreas and/or Vietnams; but this must be done in a Democratic way, much as Germany has been reunited after the Cold War.

The Communists of North Vietnam violated the Paris Peace Treaty of 1973 with the United States and attacked and destroyed the Republic of South Vietnam in 1975. Meanwhile the US congress and people had tired of the war and cut off military aid and funding to South Vietnam due to the incorrect strategy utilized by Lyndon Johnson. The US Congress also cut off funding for Cambodia; which led to the Khmer Rouge takeover of that country and the killing of millions. Many South

Vietnamese were put into Gulags or "Re-Education" camps after the communist takeover for collaborating with US forces.

The author of the book makes no excuses about the format and/or style of the book. I wanted to get my strategy published as fast as possible. Any prettying up of the book can be done later via a 2nd edition. The worst part of the Correct Strategy for the Vietnam War; was that I knew this to be the correct strategy in 1967; however I didn't have the rank to explain it to anybody of importance. I also have a dated map of the period showing my strategy from 1968; however I have miss-laid my map and so that will have to be published in a 2nd edition later. This is a "Bucket List" Book so I'm not too concerned if it is a best seller or not. It will probably be more appreciated after I am gone.....much as an artist's works are more revered after his death. Today we read Thucydides and his Peloponnesian War and wonder about him. Tomorrow; in 1500 years, someone may read my book and wonder about me. As to anyone saying I am wrong with my strategy; they should remember what one Vet once said...."If you weren't There.....Your Opinion Doesn't Count!!"

Monument to the US War Dead and Wounded and POWs

Stranger That Passes By.......

Go Tell the Americans.....

That Here;

In Obedience To Their Laws,

We Lie

This Book is also Dedicated to the 1/3rd of POW's who were not repatriated at the end of the war via the 1973 Treaty. Communist countries like to keep some prisoners as 'Bargaining Chips' after a war and papers found in the Moscow archives detail how only 2/3rds of the US prisoners of war were repatriated by The Communists. The other 1/3rd of the POWs were kept in the Gulags of North Vietnam, China, and Russia. They also include the prisoners captured on secret Missions in Cambodia and Laos; who were never acknowledged by the US Government at the time and thus could not be exchanged. Since many Vietnam Veterans are still alive; it follows that many of the POWS are still alive. Finally the book is dedicated to our Republic of Vietnam allies; the ARVN Army, Air Force, Rangers, and Marines who fought the Communists of North Vietnam so long and with such dedication; as well as our Australian and South Korean forces and other allies who fought alongside of us and bled with us.

Contents

Did Russia Win World War II?? The answer NO.

Take Britain, Canada, Australia, and the United States out of the war And take Lend lease out of the war. It is assumed that after the fall of France and during the Blitz against London the British sue for peace.....the Germans purposely delay this until the day before the Russian invasion; so Russia still thinks the war with Britain is going on. With Britain out of the war; Canada and Australia end their participation also. The US Decides to stay out of the European conflict and concentrates against Japan. Germany does not declare war on the US. This leaves the Russians with no lend lease and no allies or 2nd fronts in Africa or Italy or France.

Then Germany gets the equivalent of two more panzer armies on the eastern front. (Panzer Army "Africa Corps" and the Panzer Army left in France through 1944). It can take Germany until 1943 to take both Moscow and Leningrad; but after that the Soviet Union Disintegrates. The Soviet Rail and Road net runs through Moscow in the 1940s; so its fall severely limits Russian Mobility. After the fall of Moscow; this allows the Germans to concentrate first Against Leningrad and take that city and then concentrate to the South and take the Caucasian oil fields.

Also with two more Panzer armies; the Germans can simultaneously attack both on the Southern front in 1942 as they did historically and/or launch a 2nd Moscow Offensive in 1942.

Or the Germans could have launched just a 2nd Moscow Offensive in 1942. Result...without lend lease or allies the Russians lose the battle of attrition with the Germans in 1942 or 1943 and with the loss of Moscow and Leningrad and the Caucuses' oil fields; their communist government collapses and a "white" government takes over and sues for peace with the Germans.

How was the Battle of Midway actually won?

The US Torpedo planes were sent off alone without fighter escort which was 'delayed'; so this maneuver had the effect of pulling all the Japanese zero fighter combat air patrol planes down to sea level in order to destroy the torpedo planes. When the US Dive Bombers came over the Japanese carriers; then there were no Zero fighters on CAP to interfere with them and three Japanese carriers were set ablaze and non-operational.

The US Carriers didn't carry as large a complement of Torpedo planes as they did Dive Bombers which would seem to indicate that they were not planning on the Torpedo planes having a Decisive role in the battle; except as decoys.

This was because even though the US Navy had secured tactical and strategic surprise at the battle of Midway by breaking the Japanese naval code and stationing their carriers to the northwest of Midway in order to ambush the Japanese carriers; there was still the fact that the US Dive Bombers might not have been able to get through the cap of the four Japanese fleet carriers to do decisive damage to them in an attack.

The US; in launching a standard attack, might have only been able to destroy only one or two of of the Japanese carriers. This would have allowed the remaining two enemy carriers to launch a counter attack against the US carriers. With the US only having three attack aircraft carriers in the whole of the pacific theater ; the loss of two of these would be catastrophic and would have prolonged the war.

So; in deciding on the tactics of the battle; It was seemingly decided to send the Torpedo planes off alone without fighter escort; to draw off the enemy fighters down to sea level and thus allow the dive bombers ; coming over at altitude, to catch the Japanese carriers without a CAP over them! In this way the US achieved an astounding victory ; by setting three Japanese carriers ablaze in the first strike of the dive bombers. With only a single carrier left able to conduct air operations after the first attack; the Japanese were not able to launch a strong enough counter-attack against the US carrier fleet.

The Correct Military Strategy for the Vietnam War!!

Lyndon Johnson, His Cabinet members, the American General Staff and Officers in the Pentagon, the Joint Chiefs, The Generals in charge in the field ; all didn't know what they were doing in Vietnam in 1965 to 1975. Or; if any of the officers did; they didn't articulate the information to the President. They all; apparently, lacked Coup D'Oeil as Napoleon would have called it.....or the ability to size up a military situation at a glance.

I did; because I was already a student of military history and a Strategical Analyst as well as a Strategic Gamer. Thus I 'knew' what to do in Vietnam from the beginning. I have a plan of campaign drawn up and signed on a National Geographic™ map of Vietnam from 1968. The US violated many of the military 'maxims' of war from the beginning in Vietnam.

The first of these; as Sun Tzu would say, was always to carefully consider the decision to go to war and/or into a military action. The US Pentagon and the Joint Chiefs; as well as the eventual commanders in the field; should have been able to tell Johnson from the start the correct strategy for winning the war. For some reason; they did not do so . Johnson was famous for asking many officers and men back from Vietnam....What he should do to win the war. He should have known his strategy from the very start; before sending any troops ashore.

In war; one does not "hold the line"; nor does one seek to cause the enemy to be "fatigued"; nor does one seek to draw the war out into a "long-winded affair". Nor does one go into a war without a strategic plan for winning it. The objectives of the Johnson administration was seemingly to "fatigue" the enemy by gradually increasing forces to show him that he could not win; while at the same time, Johnson had no strategic grasp or strategy for how to win the war. Johnson's Objectives and lack of a strategy led to excessive casualties, and ended up fatiguing the US army and country. Eventually it was the US that become 'fatigued' and ended up losing the war due to public opinion. The "Search and Destroy" missions of the US forces also caused excessive US casualties....as the enemy could pick and choose which operations he wanted to ambush and attempt to overrun.

The second maxim of war is to always act to completely disable the enemy from the start. Your strategy should seek to overwhelm the enemy; to cause such a loss that he can see no way to continue the war without excessive casualties. Thus your strategy is easy to discern from this approach. I will tell you the Correct Strategy for the Vietnam War a little later.

The third maxim of war is always to strike with as much force as you can muster at once.

If your analysis is that you don't have enough troops or material or production; then according to the First maxim of war....You don't go to war. The prime example of this is in WWII when the Japanese decided to go to war with the US. Eventually the US was turning out one ferry carrier every month and eight B-24 bombers a day; which was more production than Japan could ever hope to match. The Japanese; no matter how many early victories they had, were simply over-whelmed later. Thus the Japanese should never have gone to war with the US in WWII. Neither Johnson nor the joint chiefs did an analysis of how many troops would be needed in Vietnam. And the commanders in the field evidently did no immediate analysis of how many troops they would need to successfully prosecute the war. The Vietnam War was started without any overall strategy for winning it and no adequate analysis of how many troops would be needed to execute the strategy. In Vietnam It was as if the three stooges were given command of the United States forces to fight a war.

Even if a President of the United States lacks any concepts of Military Strategy and the Secretary of Defense the same; then it falls on the selected General commanding in the field to give them instruction or lessons in what will be required. This, evidently, they did not do. It also falls upon the Joint Chiefs to do so and/or the Pentagon General Officers and staff. But we see; from the history of the Vietnam War, that none of these organizations attempted to give the President and Secretary of Defense instructions or lessons in what would be required.

In addition; nowhere is there any description of the Vietnam conflict being war-gamed before the introduction of troops to see what our forces could do and what the enemy might do.

It is instructive to remember that even during WWII and before the D-Day landing; the Germans were conducting a war game about an allied landing in France!! Thus the Pentagon of 1964 was even deficient as compared to the German General Staff of 1944 !?

An informal maxim of war would be "Do not hobble your troops; by giving them artificial constraints, which impede their freedom of maneuver" This Johnson violated by respecting the neutrality of Laos and Cambodia throughout the war and ordering US troops not to violate it. The Viet Cong and the North Vietnamese Army regularly violated the neutrality of these countries during the war. They occupied both countries and used them as a "superhighway" for bringing their supplies, troops, and ammunition into South Vietnam. The North Vietnamese practically built a "Pennsylvania Turnpike™" down through Laos and Cambodia to Truck supplies, men and ammunition into South Vietnam and they even had "Howard Johnson™" rest stops for their Truckers along the way.

Thus Johnson went into Vietnam without a Strategy, without any concept of the troop level necessary to be successful, and with hobbling the troop's freedom of maneuver before they even got there. This was a prescription for disaster; if there ever was one!

Eventually the US deployed all or parts of 15 army and marine divisions in Vietnam; when troop levels had reached their highest point.

It is clear ; from the history of the war, that a large majority of these troops should have been sent immediately to Vietnam once we had decided to go in. No one wins a war by sending in troops piecemeal .

With 10 Divisions; we can now implement the Correct Military Strategy for the Vietnam War. As shown on the map enclosed; the troops entering South Vietnam are immediately deployed along the Demilitarized Zone between North Vietnam and South Vietnam as delineated in the Paris peace conference of 1954. In addition; troops fan out into Northern Laos and are extended to the Mekong River and the Border with Thailand. The Mekong River is a large River at this point as wide as the Mississippi River in the US. All these troops dig in and produce fortifications like those of the Western Front in World War I. This blocks any North Vietnam infiltration of supplies or units from the North Vietnamese army from getting into South Vietnam and resupplying the Viet Cong. It is impossible for the NVA to supply the Viet Cong from their eastern seacoast because of the US Navy. With the line fortified; it would prove impossible for the NVA to get through Fortifications, concertina wire, mines, machine guns, mortars and artillery fire. About the only thing the NVA can do here is try to resupply the south by crossing the Mekong River twice; above and below the US fortified line; however this is a tenuous supply effort as the enemy would be subject to interdiction by US Riverine forces as well as interdiction by the US air force each time crossing the Mekong River. It also prevents The NVA from running trucks south to resupply the Viet Cong and the NVA divisions in the south.

One Marine Division, one Air Cavalry Division, and two Airborne Divisions are deployed as Reserves for the DMZ Fortified Line. The 3th

Armored Division holds Saigon. Other unassigned Brigades and units are deployed to hold various Vietnam cities and to assist the Republic of South Vietnam army units to defend against the Viet Cong. As the fortifications and development of the fortified line progressed and got stronger; some of the divisions allocated to it would be freed up and they could be used as local reserves and/or to reinforce other positions in the rest of South Vietnam.

With no supplies getting south; coupled with other strategies of the US; the Viet Cong would be unable to launch operations because of a lack of food and ammunition. And thus a buildup of the Army of the Republic of Vietnam; coupled with US reserve units would eventually destroy the Viet Cong and 'winkle out' any of their remaining elements. The enemy might also try to bull their way through the fortified Line....However that is what B-52 strikes are for.

And just as the Air Force prevented the NVA from getting a Dien Bien Phu at Khe Sanh; so they would be unable to break through this line with any consistency or with any ability to get supplies south.

China would have no pretext to come into the war on the side of the North Vietnamese; as we would not be threatening an invasion of North Vietnam when we utilized this strategy.

Thus; there would be a Free South Vietnam south of the DMZ to this day; much as there is a Free South Korea today south of the DMZ in Korea.

[encl. Map of the Republic of South Vietnam with the Fortified line along the DMZ and extended into Laos to the Mekong River with the Divisions used to defend it. These are the 5th Marines, 1st Infantry, 4th Infantry, 5th Infantry, 7th Infantry, 9th Infantry, 23rd Infantry, 24 Infantry, 25th Infantry, and 3rd Marines. In Reserve are the 1st Marines, 1st Cavalry Airmobile, 82nd Airborne, and 101st Airborne Divisions. The 3rd Armored Division holds Saigon.

Extra divisions; which had to be sent from the states are the 3rd Armored, and 7th and 24th Infantry Divisions. All the other Divisions served in Vietnam. Unattached Brigades and Regiments that served in Vietnam are detached to hold significant cities and positions within South Vietnam and to aid the Army of the Republic of South Vietnam.

Map is not to scale; but encompasses the roughly 100 miles of the DMZ in Vietnam extended through 200 miles of Laos to just above the City of Savannakhet in Laos. This is to deny the enemy the uses of the Ho Chi Minh trail and would force him to cross the Mekong River twice to try to supply the Viet Cong in the Republic of South Vietnam . This would be nearly impossible for him; since he would be denied the use of trucks to supply his forces in the south ; plus he would be under fire from air forces and Riverine craft in the Mekong River. With his supply line effectively cut; the North Vietnamese would not be able to disturb a free South Vietnam until the South Vietnam ARVN units could take over the Fortified Line for themselves.]

Correct Military Strategy of the Vietnam War Map

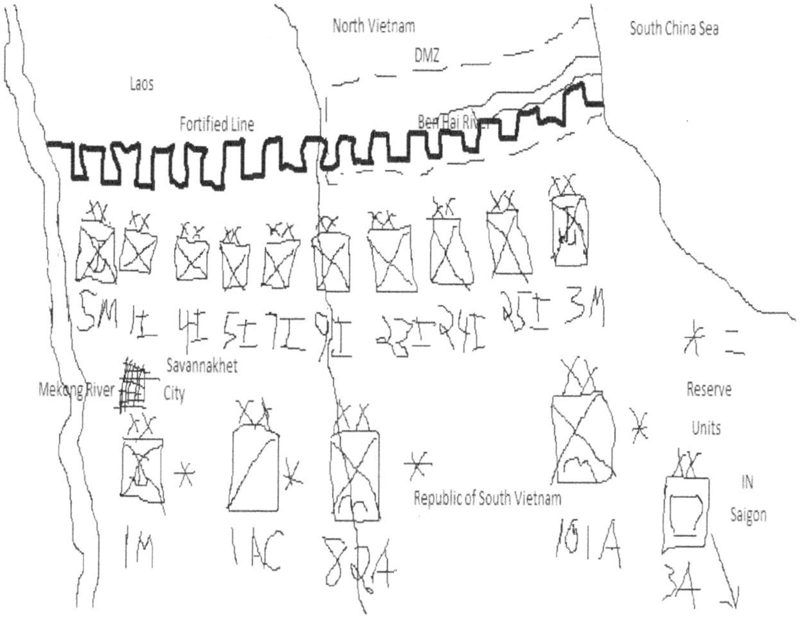

Fortified Line anchored on the Mekong River above the Laotian City of Savannakhet and extended over through Laos and Vietnam via the DMZ to the Pacific Ocean.

Fortified Line Division Units

5 Marine, 1 Infantry, 4 Infantry, 5 Infantry, 7 Infantry, 9 Infantry, 23 Infantry, 24 Infantry, 25 Infantry, and 3 Marine.

Division Units in Reserve

1 Marine, 82 Airborne, 101 Airborne, 1 Air Cavalry.

Division Unit Holding Saigon 3rd Armored.

Legend For the Correct Military Strategy For the Vietnam War Map

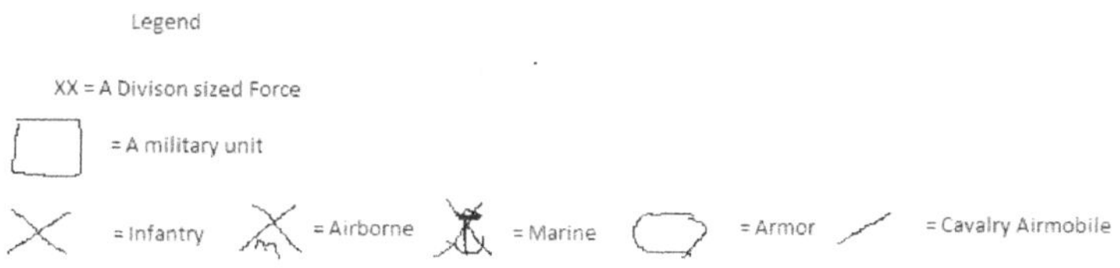

Square = a Military Unit, xx = A Division Sized Unit, Type of Unit inside Square is X = Infantry, X + Bird = Airborne, X + Anchor = Marine, [__] = Armor, / = Air Cavalry.

What scientific event could have caused the Civil War in the United States...?? Answer:-The Sunspots and Solar Flare of August 28 through September 3, 1859 .

The Earth was in the direct way of Gigantic Solar Flares which created exceptional aurora borealis and permitted the reading of Newspapers at night. These flares not only fried the telegraph lines rendering many of them unusable but also I propose People's Brains so that they could no longer negotiate or reason and/or their ability to do so was reduced. Thus; unable to compromise; as had been done in the past, a Civil War was inevitable in the United States. The extensive aurora borealis and interference with the telegraph lines was called the Carrington Effect; after the Astronomer.

With reduced ability to reason or compromise; people were fractured as to the candidates of the November 6, 1860 presidential election and were unable to accept the result of the election....Abraham Lincoln as President. Thus the South seceded and started the Civil War in the US. Instead of being able to compromise; as had been done in the past, people were psychologically incapable of compromise due to the effects of the solar flares on their brains and the south seceded after the election of Lincoln; which they saw as being unacceptable. To Date no scientific study has been attempted on the effect of solar flares upon the human brain; because many flares are not shot in the direction of the earth and because only a few massive flares have been recorded in history.

Could the Luftwaffe Have Won the Battle of Britain?? YES.

It is better to destroy an enemy air force on the ground.

The Germans should have pursued a policy of destroying the RAF on the ground. This is achieved by sending 'fake' bombing raids against the British Cities; And waiting for the Spitfires and Hurricanes to be launched to oppose them and eventually run out of gas. The British would have sent their fighters up to stop the bombing raids on the cities while the Germans would have launched real raids on the RAF airfields the British planes would have to land at to be refueled.

Of Course this requires a close estimate of the gas consumption of both a spitfire and a Hurricane which knowledge would have had to be stolen by a spy or learned by disabling a British aircraft and forcing it to land in France.

The bombing raids against the British airfields would be launched; perfectly timed to be over the airfields when the British planes are refueling. In this way most of the RAF is destroyed on the ground in 2 to 3 weeks . Since the RAF at the time consisted of less than a 1,000 fighters; this German strategy would have decimated those elements and would have enabled the German forces to have launched their Operation Sea lion to land an army in England.

England might have adopted a strategy of converting Bomber command and its Bombers into kamikazes; but the invasion would have been supported by the German fighters, so this desperate effort would not have worked.

Was Pickett's Charge really a Charge?? Partly.

In my estimation it was both a Charge and a Demonstration Maneuver designed to crack Union army morale and panic and rout them.

Lee was Overconfident. After a year of commanding the Army of Northern Virginia; Lee had basically driven the Union army away from Richmond in the Seven Days, Had routed Pope at Second Bull Run, and had invaded the north at Antietam. Then he had repulsed the Union Army attack at Fredericksburg and then had routed the Union army at the battle of Chancellorsville. In Lee's mind these victories were not entirely due to the poor generalship of the union army; but were also due to the poor fighting ability and low morale of the union army Soldiers.

Thus, At Gettysburg; after two days of Fighting Lee thought that the Union army Corps were in low morale; having routed the Union I, XI, and III Corps; as well as severely handling elements of the union XII, V and VI Corps. Lee thought that Union soldier morale was low and an impressive assault on their center would cause them to panic and rout. Lee also thought that Meade had reduced the forces in his center In order to reinforce his flanks and thus the center of the union line was weak. In Lee's mind the appearance of a Full Corps assault; three divisions of the Confederate army, would demoralize and panic the union army which would break and run from the battlefield.

So you see; in Lee's thinking, the assault was not only a charge; but was a demonstration maneuver which was designed to panic and rout the Union army. Longstreet's Opinion that no troops ever arrayed for battle could take the union center was thus besides the point in Lee's mind.

Did General Sickles's advancement of his IIIrd Corps at the battle of Gettysburg Save or almost lead to the destruction of the Army of the Potomac ? Answer...Saved...

Sickle's advancement of his IIIrd Corp's 2 divisions allowed the Confederates to Contact the Union army without running a gauntlet of artillery fire like Pickett's charge; so Sickles aided the Confederates. On the other hand it delayed the Confederates from pushing deep into the flank of the union army and threatening or seizing the Round Tops Earlier than they did.....so Sickles actually saved the Union army at Gettysburg by preventing their left flank from being turned.

The time the Confederates used up and the configuration of the forces they used to reduce the 'V' formation of the IIIrd Corps prevented the Confederates from extending their battle line to the south and; in turn, prevented them from further outflanking the union army on their left flank.

George Meade; the commanding officer of the Army of the Potomac, had told Sickles to stay in the line of battle of the army extending from Cemetery Ridge to the Round Top Hills. The Union army was engaged in a desperate battle and perhaps the premier battle of the Civil War....A Loss here; with the army routed back to Washington, might lead to European recognition and assistance to the Confederacy.

Technically; Sickles had blatantly disobeyed Meade on the field of battle; a fact; which; in many other armies, would have got him court-martialed and cashiered, and/or shot out of hand by Napoleon. However; in the aftermath of the Victory and because Sickles was a 'Political' general; no action was taken against him. General Sickles ; who had lost a leg in the battle, was told that the War Department felt that he was too disabled to command troops in the field and thus had to retire.

The second day at Gettysburg was actually the day the union army was in the greatest peril of having it's Left flank turned and this might have routed the army. Thus; I have to come down in favor of General Sickles; who blunted and occupied the Confederates and; in a way, prevented them from extending their line of battle and out-flanking the Union army.

Sickles advanced merely to occupy a hill in his front to prevent the rebels from placing artillery on it to work his corps over. Thus Sickles; by disobeying orders, saved the union Army of the Potomac and the United States at Gettysburg , by mistake. As some wit once Said……A higher power watches over Fools, Drunks, and the United States of America!!

Was Pearl Harbor Japan's Greatest Victory in WWII or its Greatest Failure?? FAILURE.

Japan should have invaded Hawaii after the attacks to take the US naval base. This would have forced the United States to start all campaigns out of San Francisco or San Diego. The Japanese operation against Midway in 1942 shows that they should have invaded Hawaii at the beginning of the war; which would have made the Midway operation unnecessary!

If the Japanese had invaded Hawaii; they would have captured the US battleships and carriers and fleet intact ......thus lengthening the war at least a year or two and it would have given them a great advantage in the Pacific; as they could have interdicted the supply lines to Australia and the South Seas Islands. Australia might have fallen to another Japanese army invasion. An invasion of Hawaii would have resulted in the Japanese seizing the battleships and carriers of the US Pacific fleet; which would have immeasurably strengthened the Japanese navy while seriously weakening the US Navy. The Japanese launched many successful amphibious assaults during the beginning of WWII; so there would be no reason why they couldn't have launched enough transports to ferry several army divisions to land in the Hawaiian islands; simultaneous to the carrier attacks to destroy the US planes and airfields and defensive positions. The Midway operation had included transports for ferrying invading troops to Midway Island.

Failing to try to invade the Hawaiian Islands; the next best strategy for the Japanese was to launch a second wave of carrier air strikes against the remaining battleships and support ships and to severely damage the other naval facilities in Pearl Harbor. But the Japanese admiral didn't launch even the additional air strikes; but took his carrier fleet back to Japan without causing additional damage to the remaining ships of the US Navy.

Thus the Japanese strategy of attacking the US Fleet at Pearl Harbor and Hoping that the United States would sue for peace in the wake of the disaster was a strategy predictably doomed to failure. And the US usually enforces 'Unconditional Surrender' on its opponents once a war starts and fights through to that conclusion. Thus General Grant's nickname; based on his initials. U.S. or "Unconditional Surrender " Grant!

Did Lincoln Pick Good Generals for the Army of the Potomac?? NO.

Lincoln usually picked very bad generals for the army of the Potomac.

McDowell moved too soon against the Confederates and without enough troops. The Confederates concentrated against him by bringing the Confederate army in the Shenandoah Valley down by train to reinforce their army in Manassas.

McClellan; even though he was young and popular with the soldiers, always thought that the Confederates outnumbered him 2 to 1 and never wanted to engage them because of it. He allowed Lee to drive him away from Richmond in the Seven days battles.

Pope; even though he took Island Number 10 in the Mississippi River was Incompetent and let the army get flanked at second Bull Run.

Burnside; even though he had launched a partly successful flank attack on the Confederate army at Antietam, was not a professional soldier and ended up foolishly launching the Army of the Potomac in a useless headlong assault at Fredericksburg; after the Confederates had plenty of time to prepare their defensive positions.

Hooker; even though he was an intelligent commander, lost his nerve against Lee in the middle of the battle got drunk and then couldn't maneuver and ended up getting the Union army flanked at Chancellorsville by Jackson's II Confederate Corp.

Meade; who apparently was picked because he wasn't eligible for the Presidency, was too hesitant and allowed Lee to steal marches on him. Meade never engaged the Confederate army in an attack after the battle of Gettysburg. Also Meade missed the fact that the Confederates had sent the First Confederate Corps to Georgia to reinforce Bragg for The battle of Chickamauga in September of 1863. Even after learning of this fact; Meade did not immediately engage the Confederate army under Lee. From July, 1863 until May of 1864 the Army of the Potomac did nothing under Meade except await the arrival of Grant. By contrast; when the union transferred the XI and XII corps to the western theater to aid the union army besieged in Chattanooga, Lee immediately launched an offensive around the flank of the union army towards Washington which Meade had to respond to.

Grant was an obvious choice to command all of the armies of the US by 1864 because he had won virtually all his engagements against the Confederates; with only Shiloh as a Draw. So Lincoln picking Grant to command all of the Union Armies in 1864 was a No Brainer.

Who Was the Confederacy's Best General??

Lee and Jackson are the Generals usually put forward as the south's best Officers. However; also in the running is Pierre Beauregard.

Beauregard won the First battle of Bull Run.

And Beauregard participated in the planning and attack at Shiloh; which nearly did Grant's army in, in 1862 and also reduced Lincoln's confidence in Grant early in the war or Grant might have been transferred to command the Army of the Potomac sooner.

Later in the war; Beauregard saved Petersburg; when he telegraphed Lee to send troops to Petersburg after discovering that Grant had sent a corps of the union army south of the James River. Thus Beauregard saved the Confederacy for another Year.

Beauregard was under-rated as a Confederate general by Jefferson Davis and/or was on the political outs with Jefferson Davis otherwise he should probably have commanded the Army of the Tennessee instead of Bragg or Hood and/or should have commanded the Confederate forces around Vicksburg against Grant instead of Pemberton. Beauregard would have undoubtedly secured the surrender of the union army at Chattanooga; where it was routed to , after the battle of Chickamauga. Bragg inexplicably let this golden opportunity slip away; which would have been the largest surrender of Union troops in the civil war.

Did the US use the Correct Military Strategy in the Second Gulf war?? YES.

Instead of attacking with all divisions; like in World War II, the US attacked with only one division at first; because they were afraid that Saddam would hit the lead division with SCUD Missiles full of Chemical weapons; which would have disabled it.

If this had happened; the divisions in the rear would have bypassed the disabled division and kept going....Thus, it was the correct decision to attack with only one division in the advance; instead of all divisions at once; such as one would have done in World War II.

Was Lee's Plan to join Johnston to defeat Sherman doable?? YES-However.

After the fall of Petersburg and Richmond; If Lee could have marched away from Grant; and joined Johnston, the Combined Confederate armies might have been able to defeat Sherman and drive him back; however Sherman could just have retreated to Wilmington and the cover of the Union fleet guns. Grant would have rapidly advanced on their rear and the Confederates would have sooner or later faced the combined Grant-Sherman Union armies. So an attack on Sherman would have merely delayed the ultimate Confederate surrender by a Few months.

Could Napoleon have won the Battle of Waterloo?? YES.

Without Blucher coming up to support Wellington; Napoleon had the three divisions of the Imperial Guard in reserve plus two divisions of the VI Corps in reserve. These would have been used to thrust at Wellington's Center; late in the battle which would have had a good chance at breaking through it and then routing the British. Losing the battle; the British probably would have evacuated their army from the continent at Antwerp. This would have enabled Napoleon to turn again against the German army; who would have lost again without their English allies.

With Napoleon having defeated both the English and German armies; the Russians and the Austrians would not have been so ready to march to engage Napoleon. This would give him additional time to raise new forces and negotiate to stay in power in France.

Was Glasnost and Perestroika a KGB plot to prevent the US from putting up SDI?? MAYBE.

Because they had lost the Cold War to the United States and couldn't match the technological advances of the United States; the Soviet Union under Gorbachev tried everything to dissuade President Ronald Reagan to stop the United States Deployment of an ICBM Defense called the Strategic Defense Initiative . When these efforts failed in the Iceland summit meeting; the Soviets were left with only 3 choices; as they saw it......

1. Surrender; Out of the Question; since it confirmed their loss in the Cold War.

2. Immediate ICBM surprise attack on the United States; still out of the Question, since it abrogated their strategy of eventually perverting the West into an adoption of Communism and/or surrender and also because this would have resulted in the destruction of Russia.

3. Or a Strategy of Subterfuge.....!! The KGB had come up with the masterful plan to fake the fall of the Soviet Union and adopted the fiction that a coup under Boris Yeltsin had taken over from Gorbachev. In this way they would Lull the United States into a false sense of security and then opponents of Reagan's SDI could parrot the Russian line and say that it was no longer necessary. The real control of the important infrastructure of the Soviet Union ; such as the ICBM nuclear forces and armies and Special Forces and key industries and political

control structures would be kept under the control of loyal KGB operatives. The United States under Bill Clinton seemingly fell for this tactic and ceased work on putting up an SDI defense of the United States; so the KGB Ruse was successful and had staved off Reagan's SDI.

Later the KGB could come back out into the open and take over Russia again and then re-establish the old Soviet Union; once the United States had failed to put up SDI .

So the KGB adopted option 3; even if it meant the loss of the Warsaw Pact nations and the temporary independence of other parts of the old Soviet Union. After a period of time; the KGB planned on re-establishing the old Soviet Union ; with themselves still in charge and meanwhile the United States would not have deployed their Strategic Defense Initiative against them to render their ICBMs useless.

This has seemingly happened with the rise of Vladimir Putin as Russia's President...an ex- KGB officer whose family has been Communist since the revolution in 1917 !! The Soviets have had time to bring their own ICBM defenses up to snuff as well as develop newer ICBMs and ICBM Supersonic Rocket Motors as well as newer Cruise Missiles and Stealth Submarines. Significant ICBM and atomic bomb support has been lent to states antagonistic to the US such as North Korea and Iran. These are basically new allies of the new Soviet Union.

The KGB maneuver has thus been eminently successful. The United States strategic ICBM nuclear forces in the mean time have been in decline; with no new ICBMs or their rocket motors and/or Supersonic ICBMs or Cruise Missiles being developed.

Would Reagan's SDI have worked to defend the US against ICBM attack?? YES.

You can "Hit a Bullet with a Bullet" as NASA itself has demonstrated repeatedly In intercepting, orbiting, impacting and landing on various Asteroids and Comets and in the Fly-Bys of all the Planets including Pluto. Space satellites have not only completed their first missions lately; but have gone on to Intercept other asteroids and planetoids in the same mission.

Reagan's SDI could work to defeat the attack of Land Based ICBMs and Cruise Missiles By using a Shot-Gun type approach; the defending ABMs don't have to differentiate between Actual ICBMs and Fakes....since they will be blanketing and hitting all missiles including fakes With a Shotgun pellet type approach to hit anything incoming. The ABMs can be fitted with a System which detects when they are within range of incoming missiles and then they would fire their 'shotgun pellets' to disable the missiles. This is one possible type of defensive weapon against the ICBMs and Cruise Missiles .

In the near future; Laser weapons should allow an even greater range of accuracy to hit the ICBMs and Cruise Missiles before they can close with you.

As a Defense against ballistic Missile submarines; the United States will have to develop a kind of anti-sub mine defense around the north America continent to defend against stealth enemy ICBM subs sneaking next to the continent to launch their ICBMs with shorter times of trajectory to impact so as to overwhelm the defenses. This will have to be a new system of anti-submarine mines so as to disable enemy stealth submarines before they can get close to the United States to launch their ICBMs and/or Cruise Missiles. The system of anti-submarine mines would be capable of being turned on and off; so it would not target subs or ships in peacetime.

The United States is currently in a state of extreme peril in regard to nuclear weapons; as it is obvious to any observer that both Russia and Communist China are furiously developing First Strike Nuclear ICBMs and Cruise Missiles. These are Hypersonic ICBMs and Cruise Missiles as well as specialty ICBMs and Cruise Missiles designed to confuse and outwit any defensive ABM defensives. Moreover; they would not be developing such first strike weapons unless their own strategic defense ABM system had been developed and improved. The United States meanwhile has lacked the Will or strategic direction to develop Newer ICBMS and Rocket Motors as well as newer Cruise Missiles; such as Hypersonic ones. In addition a lassez- faire attitude has prevailed with the United States lacking the will or strategic direction to develop its own strategic defense initiative of an ABM and cruise missile defense. Thus my warning is that the United States is just asking to get nuclear bombed!!?? In addition; both Russia and China have aided in the development of nuclear ICBMs and atomic bombs for client states antagonistic to the United States such as North Korea and Iran. Thus there could exist a four way nuclear alliance against the United States with Russia or China using the client states to attack first so as to confuse and distract the United States defenses and counter attacks.

If he had lived; would Genghis Khan have conquered the World?? YES.

The Mongols campaigned even in winter; when European armies were bogged Down...The Mongols carried their rations with them in the form of beef jerky thus they were not subject to the same supply problems of the infantry of the kingdoms in Europe. Genghis had already conquered all the fortified cities of China...so those in Europe could not have held out against him.

Genhis's sons and Generals destroyed the Polish army at Liegnitz in April 1241 and another Hungarian army at Mohi in April of 1241 as a precursor to their invasion of Europe the next year.

The European armies were hastily thrown together amalgams of forces; many of them untrained, compared to the Mongol armies; which were well-trained and veterans of many a campaign. Thus Europe was ripe for a Mongol Invasion and could not have held out against him. Fortunately, Genghis died relatively early in his 60's and the Mongols always returned to Mongolia to elect a new Khan; so in this way Europe was frequently spared.

What was the Confederate Strategy in 1863??

The Confederates debated sending a Confederate Corps from the army of Northern Virginia; probably Longstreet's First Corps, to drive Grant away from Vicksburg (as they would do later in the Chickamauga campaign in Georgia)....but they recognized that this would only be a temporary solution.

They decided instead to reinforce Lee; who they considered their best General, with all available troops and have him seek a decisive battle in the north in either Maryland or Pennsylvania which might lead Britain or France to recognize the Confederacy; particularly If the Confederates succeeded in capturing Washington , DC after the battle. Since the Confederate stronghold in Vicksburg could hopefully hold out a while......it would give Lee the time to win a decisive battle in the east; and then the fall of Vicksburg would then seem irrelevant . Vicksburg did hold out until July 4th, 1863; and by that time Lee had already been repulsed at Gettysburg on July 3rd.

Could the Confederates have captured the Union army in Chattanooga?? YES.

The Confederates should have done all they could to capture that army; which would have been a decisive victory which might have led to European recognition of the Confederacy.

Longstreet's corps should have been kept in the area longer to interdict the union supply lines to Chattanooga; instead of starting the Knoxville campaign.....which was to start to move Longstreet's Corp back to the Army of Northern Virginia. General Meade with the union army of the Potomac was doing nothing at this time to threaten Richmond; so there was no need for Longstreet to start back right away.

It is one of the Mysteries of the Civil War....Why the Confederates did not pull out all the stops to capture the Union army in Chattanooga....after their victory at Chickamauga...!!??

Beauregard would have been a better commander for the Confederate Army of the Tennessee than Bragg. The Confederate commanders wasted valuable time arguing among themselves and a cadre of officers wanted to have Bragg relieved of command even after the victory at Chickamauga.

The Confederates were overconfident as they regarded the Union army of the Cumberland as almost finished and incapable of offensive attack against them. Thus it was that Bragg failed to properly fortify Missionary Ridge and Lookout Mountain; both of which were lost to union assaults rather easily. Missionary Ridge in particular being a "soldier's battle" where the troops having halted at rifle pits below the ridge; found the Confederate artillery fire so troublesome that they assaulted up the hill on their own. If Bragg had properly fortified the Ridge; the union troops should have been tied down under fire in a kind of repeat of the battle of Fredericksburg in Virginia. As it was; after a short sharp battle the Confederate army was routed and the battle of Chattanooga was won for the union. Bragg was sacked after the battle and replaced by Johnston.

Did Lincoln overestimate the Union armies ability to pursue Lee's army after a desperateBattle? YES.

Lincoln chastised McClellan for not pursuing the Confederate army after Antietam which it was incapable of doing after assaulting the Confederate army all day in September of 1862.

And Lincoln chastised Meade for not pursuing and destroying the Confederate army after Three days of desperate defensive battle at Gettysburg in July of 1863.

Lincoln consistently overestimated the Union troop's ability to pursue after fighting a desperate battle.

Future Strategical Considerations for the United States (as of 2018)........

West Point, Annapolis, and Boulder....the Service Academies; In order to produce officers having Coup D' Oeil, it will be necessary for them to adopt classes in military or strategical gaming in all four years of their Curriculums. This is the only way for officers to develop this; short of being in an actual war. This should not be left only for the development of senior officers; as Vietnam has shown that junior officers; who had been thrust into command of units in the war; lacked military experience or the Coup D' Oeil.

Officers and Generals having Coup D' Oeil can advise future Presidents of the United States on how to conduct operations in warfare and should be able to prevent any future Vietnams.

North Korea

At the present time; North Korea and Kim Jung Un are a problem for the United States because the Dictator is always threatening to attack the United States and he is bent on developing ICBMs and atomic bombs. While many Presidential administrations have tried to negotiate with Kim to ease this threat ......all have failed.

Unless Kim Jung Un agrees to disband his ICBMs, Cruise Missiles, atomic bombs, and satellites and demobilizes most of his army and agrees to a peace treaty for the Korean War; or Kim Jung Un is removed from power; it leaves the United States with no alternative but to eliminate Kim from the world stage. If this results in a local war with North Korea; then that is the cost of ensuring world peace.

The United States should try to take Kim 'out' first by any means necessary; such as Cruise Missiles, Drones, and/or Bunker Buster bombs. If during a Missile test it might even be configured as a North Korean missile launch failure that takes Kim out. After that; if any military reaction from North Korea; then the US Cruise Missiles and air force should take out the entire North Korean infrastructure, ICBM launch facilities and vehicles and cruise missile launch facilities as well as any other military objectives. The North Korean army and artillery can be taken out the same way including strikes from B-52 Bombers and stealth bombers.

Iran

It is clear that Kerry's nuclear treaty with Iran didn't go far enough. Since this Iranian Regime insists on supporting terrorists in many areas. The United States must move to limit Iran's involvement in the rest of the world and especially in the Middle East. It is unfortunate that Iran was able to send some military elements to Iraq to fight ISIS under Obama; since this scrambles the Middle East situation; and now they must be gotten sent home to Iran. US military forces should be increased in Iraq; in order to counteract any Iranian and Russian influence. The nuclear treaty with Iran must be renegotiated to insure that Iran does not get nuclear weapons and ICBMs and Cruise Missiles. This needs to be undertaken after Iranian troops are withdrawn from Iraq and operations against North Korea are concluded.

North Vietnam

North Vietnam should be held accountable for the POWs that they didn't return as part of the 1973 treaty ending the Vietnam War. It is well known that NV sent prisoners to both China and Russia during the Vietnam War. And papers found in the Russian archives after the fall of the Soviet Union show that NV only returned about 2/3rds of the POWs it had on the conclusion of the war. The rest it kept as part of a future bargaining 'chip'. The United States government should insist on the return of all live prisoners of war from North Vietnam. And it should insist that NV get back all of the prisoners it sent to China and Russia and turn them over to the United States. The United States should quit making sweet economic deals with North Vietnam until we get all of our POWs back.

China

The Chinese have illegally seized the South Chinese Resource area and have been developing man-made islands and militarizing them to defend it. World bodies ; such as the UN and the World Court; with the backing of the United States, should be made to condemn this blatant seizure of territory and demand that China cease militarizing the area and that the area should be open to all the local nations. Based on economic strength and the size of their armed forces; which are being modernized and increased every day; China is now the number one antagonist of the United States in the world. Their seizure of the South Seas resource area is illegal. China has been building underground sub bases and Pens for their nuclear subs such as those the Nazis had in France in WWII; and are steadily becoming a modern naval force to be reckoned with. The Chinese economy is growing stronger and the sanctions against Russia are counter-productive as this is driving Russia into becoming a Chinese ally as the Chinese can act to bolster the Russian economy against the sanctions. China has produced and put into service Hypersonic ICBMs which will make it harder for the US ABM defenses to protect the United States. Our Intelligence Services have failed to keep the various Presidents informed as to Chinese capabilities and deployment of these Missiles.

Russia

The Russians are busily modernizing their nuclear subs and tanks; as well as their ICBM strike force. They remain a formidable adversary of the United States. It is a mistake for the United States not to modernize its Rocket motors, ICBMs, and Cruise Missiles under the terms of our treaties with them. As newer ICBMs come on line the old ones can be taken out of service and defused as per the treaty missile limits. Congress and the President should work together to resolve the Ukrainian crises as well as any other differences. The current Congress hobbling of President Trump; by trying to prevent him from taking off or relaxing sanctions against Russia, is counter-productive to allowing him to resolve the Ukrainian situation and to bettering relations with Russia. Also laws passed by congress hobbling any president are actually Un-Constitutional ; since they represent an Legislative intrusion into the operations of the Executive branch. A prime example of an Un-Constitutional law of this type is the Iran-Contra law passed against President Reagan. The economic sanctions against Russia will be found to be ineffective as the Chinese can always bolster the Russian economy to have Russia as an ally.

To encourage both Russia and China to become allies together with each other is counter-productive to the interests of the United States. It would be better for the US to try to become allies with Russia as an aid against the Chinese. For this reason; President Trump should be given a freer hand to drop economic sanctions against Russia in return for an equitable settlement of granting independence and territorial integrity to the Ukraine. If Russia wants to put a pipeline across the Ukraine to sell gas to the EU; this could be approved, if they agree to giving the Ukraine it's gas for free in return for the pipeline thru Ukrainian territory.

Russia has produced and put into service Hypersonic ICBMs which will make it harder for the US ABM defenses to protect the United States. Our Intelligence Services have failed to keep the various

Presidents informed as to Russian capabilities and deployment of these Missiles.

The Arctic

Russia has been busily claiming a lot of the arctic ; in order to seize the resources in the area; such as oil reserves. The UN and the World Court should be watching this closely and should be ruling on claims in this area; with the backing of the United States. The United States and Canada must make sure that they get their fair share of the Arctic region. This is an area in which the claims should be resolved as soon as possible in order to prevent miss-understandings and possible war over the area.

The Middle East

Obama committed many strategic blunders in the Middle East; which have turned the Region into a more dangerous place and may have set up a situation leading to world war III.

First; Obama took too long a time to remove the dictator Assad in Syria; allowing the Russians to reinforce Assad and establish troops in the Middle East.

Secondly; Obama removed US troops from Iraq too soon; allowing ISIS to take over the region and also providing an excuse for Iranian troops to enter Iraq under the guise of defeating

ISIS.

These two events almost require that more US troops be sent into Iraq; in order to stabilize the situation until the Iranian and Russian troops are removed. Any dreams of removing Assad in Syria are much too late; and the US must accept the fact that he is in charge in Syria for the time being.

Russia and Iran must not be allowed to seize and/or "colonize" Iraq as this will lead to an attack upon Israel necessitating US involvement and a drift towards World war III.

Lebanon

Lebanon should be occupied by a peace-keeping army of the United Nations and all residents Dis-armed. Any Terrorist organizations should be outlawed and kicked out of the country. Outside interference in Lebanon; such as from Syria, Iraq, Iran, or Russia should not be allowed. The UN can then foster economic development in Lebanon by building up the Hotels and Casinos on the Beaches; so the residents will have good jobs and the country can become a peaceful showcase. Lebanon would be a UN Protectorate until its government and internal police and army security forces could take over from the forces of the UN. At that point the UN could withdraw.

Afghanistan

I really can't think of a strategical reason for the US to be still in Afghanistan. This only makes sense if we envision war with Iran in the future and we want to attack Iran from all sides. It does keep the terrorists from using Afghanistan as a training ground; however that seems to be an expensive proposition which also exposes our troops to casualties and keeps US forces strung out all over the globe. A Much less expensive option would be to cultivate a younger warlord to run Afghanistan supplying him with weapons and monies and air support; if he would serve to keep the Taliban and Al Qaeda out of the country. This would allow all of our forces to be withdrawn to reinforce Iraq.

Egypt

Egypt is to be Commended for adhering to the Jimmy Carter agreement for peace between them and Israel. Egypt's foreign aid and assistance should never be cut off while this continues.

Pakistan

Pakistan has done yeoman's work in combating against Al Queada and Taliban elements in their mountainous regions and in aiding the US in our war on Terror campaign and thus their foreign aid and assistance should never be cut off while this continues.

Future Alliances

Actually the United States and Russia are natural allies against both the Terrorists and the Chinese. If the Chinese ever got to where they neutralized the United States; it wouldn't be too long before they would be trying to neutralize Russia, also. The Terrorists have commited terrorist actions against both the Russians and the United States. Also; both the United States and Russia are basically Christian nations.

For these reasons; both Russia and the United States should put aside their differences and seek to become allies; in order to stabilize the world.

The Ukraine

The US Congress is short-sighted in refusing President Trump's ability to take sanctions off of Russia. Only in this way can a negotiated settlement to the situation can be had.

Russia wants to sell its gas in the EU with a pipeline through the Ukraine. This is easily negotiated. Russia guarantees the present borders and independence of the Ukraine. The Ukraine allows Russia to build a pipeline across its territory. Russia pays the Ukrainians by giving them free gas.

The Internet Cyber Warfare.

The Servers and Internet of the US government and US Corporations and Companies must be made secure against the threat of enemy hacking , Intrusion, and spying.

This can be achieved by an Intranet-Internet strategical setup.

All US Government processing of information must be done on an Intranet; within each government agency or organization, which is not directly connected to the Internet.

The Internet connection of each government agency would take in the information from the internet and first examine the data for viruses, root kits, malware, etc. These would be eliminated from the information retrieved.

Utilizing a dual disk drive setup the information would then be passed for processing to the agencies intranet. Any responses would be passed through the dual disk drive setup back for response to the Internet.

This type of setup only inconveniences us in the type of systems which are real-time.....

However the transmission of data between the Intranet and the Internet will eventually be speeded up by chips so as to make the retrieval and re-transmission of data virtually simultaneous by means of disk drives and/or chips between the systems. The actual storage servers and databases would be kept on the Intranet and would not be directly attached to the regular Internet; so their information would not be able to be hacked.

In the same way the voting of each state in the United States can be protected by having each state do it's voting by means of an in-state Intranet; not connected to the Internet. Thus each 50 state's elections are not easily hacked or interfered with. The results of each state's voting can be carried to a central location and tabulated on a central Intranet ; say in Washington, D.C., for verification of each state's results. The results of the election could then be transmitted to the news media.

The US Electrical Grid

The United States Electrical grid must be upgraded and modernized ; so as to be able to be capable to withstand Solar Flares and enemy ICBM EMP attacks. It must not be subject to being knocked out for any length of time by these types of events. The number of self sufficient grids needs to be increased; for instance Washington , D.C. would need to be capable of functioning off of a "Virginia" grid or a "Pennsylvania" grid. A Grid should probably encompass no more than several states and/or a particular region. At present the US does not have enough stand alone electrical grids.

US Anti-Sub and Ship Defenses

The United States needs an anti-submarine mine defense ; capable of being activated in a military crises which would act against enemy ICBM and/or cruise missile subs and ships in order to prevent any enemy; such as Russia, China, North Korea, Iran, etc., from closing up ships close to our coastline to launch nuclear weapons from a shorter range. The system should be capable of being turned on and off. This is even more important with the development of Russian and Chinese Hypersonic ICBMs which may be installed on their nuclear sub fleet.

Earth Asteroid and Comet Defense.

The United States, Russia, and China; as well as other nations and the UN should all be collaborating on a Defense for the Earth against an Asteroid or Comet strike .

This is probably best done by landing nuclear or other type engines on an asteroid to give them a boost in a direction so as to miss the earth.

A destructive or explosive method is most likely counter-productive as fragments of the asteroid or comet would then still strike the earth producing just as much damage.

Since it is clear; according to reports, that many asteroids are missed in detection coming from various angles against the Earth......the earth's means of detecting intersecting asteroids or comets must be improved upon. Many more observatories are needed as well as satellite detection must be utilized.

Robot Telescopes might be placed on the face and reverse of the Moon in order to facilitate additional tracking of incoming Asteroids and Comets so that we have better advance warning of incoming asteroids and comets. At present many asteroids are narrowly missing earth and they have been detected or even missed too late to be able to move them out of colliding with the earth. This should be a major initiative of any US President's focus; once the national defense of the United States has been taken care of.

The Current World Situation for the United States

The Current situation is at a High Crisis Point for the United States as enemy nations are improving and modernizing their ICBMs while the United States does nothing and utilizes 40 year old ICBMs to defend itself with.

The United States needs to modernize its land based Nuclear ICBM arsenal immediately and also modernize the ICBMs on its nuclear submarines.

China, Russia, Iran, and North Korea are all probably in a secret military and economic alliance against the Unites States. Sanctions against Russia and North Korea will have no effect as long as the Chinese economy can keep them afloat.

It is clear that the only way that North Korea and Iran have developed ICBMs and Cruise Missiles so fast; is if a major power ; such as Russia or China, had provided them with technical assistance. For this reason; I doubt that appeals to Russia and/or China to curb North Korea will have any effect. This leaves removing Kim Jung Un by any means necessary and/or destroying the nuclear capability of North Korea the only possible solution. After fixing North Korea; then the US will have an easier time adjusting the Iranian Nuclear treaty to eliminate their ICBMs and Cruise Missiles development more to our liking.

Of Course the United States should foster peaceful relations with China , Russia, North Korea, and Iran. However; Treaties should restrict North Korea and Iran from developing ICBMs, Cruise Missiles and atomic bombs. And negotiations with China should be undertaken to try to remove their military presence from taking over the south seas oil resource area. The area should be open to all countries to develop and drill for oil. And negotiations with Russia over the Arctic should be undertaken in order to apportion the resources fairly between the nations concerned including Canada and other countries.

Negotiations with North Korea should entail the following points

China, United States, North Korea, South Korea all agree to a Peace Treaty for the Koran War.

All signatories agree not to attack the other nations in the future.

All POW s kept by the signatories are returned to their respective nations.

All signatories agree not to seek or encourage a government or regime change in the others.

The US ceases military training and war games in South Korea.

North Korea gives up all ICBM, Cruise Missile, and atomic or other nuclear bomb manufacture. And in addition; North Korea destroys its present stockpiles of ICBMs and Cruise Missiles s well as atomic bombs.

North Korea ceases to produce materials for making atomic or other nuclear bombs.

North Korea agrees to accept UN and other Nuclear Inspectors into its country in order to verify that treaty terms are carried out.

Basically the same terms should be enforced with Iran.

Epilogue

Suetonius; in his life of the Ceasars, has said that the Roman Emperor Augustas Ceasar; the Conqueror of the known world and the man to whom millions owed their lives and loyalty to, used to bang his head against the walls of his Grand Palace and scream the words "Quintili Vare, legiones redde!" which means basically "Quintilius Varus, Give me Back my Legions!". Varus had led three legions and auxiliaries into a trap across the Rhine River; betrayed by a german scout, and had got them killed to the last man, losing 20,000 men and the eagles from three Legions; A major disgrace to the Romans.

The battle ended any further Roman expansion in germany and northern europe; though some punative expeditions were launched. Lyndon Johnson lost 60,000 men in Vietnam, Laos, Cambodia, and South East Asia. That's the equivalent of six divisions of infantry at 10,000 men a Division!! And this may not include the 1/3 of the POWs which were withheld by the North Vietnamese Communists as bargaining chips. It is also known that North Vietnam gave some of these POWs to the Communist governments of Russia and China to be interned in their Gulags. Johnson had lost the Independent Republic of South Vietnam to the North Vietnamese Communists. In addition he lost Laos to the Communists and enabled the Khmer Rouge to take over Cambodia and execute millions.

And so to end the final words of my treatise on the Vietnam War; we can echo the Roman Emperor Augustas Ceasar's words...... "Lyndicus Johnsicus, Give me Back My Legions!"

About the Author

Merrell D. Wilson is a Military Historian and a Strategical Analyst and Gamer.

He served in the United States Regular Air Force in the Vietnam War.

He Has an Associate in Arts Degree and a Bachelor of Science Degree.

He was named after Major General Frank Merrill of Merrill's Marauders

Fame in World War II. His Sister; Glenna, was partly named after the B-24

Bomber; the Glenna Bea, in which a relative was a crew member of and flew in

40 missions in the Pacific in WWII.

Made in United States
Orlando, FL
05 November 2026

90230233R00049